Dr JM Field (Spearim) is a Gamilaraay mari from Moree way, but grew up on Darug land in a small town along the Great Dividing Range. He studied maths and French literature at the University of Sydney, before completing a doctorate at Balliol College, Oxford. He remains a research mathematician, with a focus on the mathematics and genetics of traditional kinship systems. He is also the author of *Etta and the Shadow Taboo*, which was highly commended for the Victorian Premier's Literary Awards (Indigenous Writing) as well as shortlisted for the Prime Minister's Literary Awards (Children's Literature).

Praise for *THE EAGLE AND THE CROW*

'During my first read of JM Field's book I had only an inkling of what he wrote. Of course I didn't – and couldn't – put it all together then. I will need to reread multiple times to get somewhere closer to understanding fully. This is something I should have read in my early teens. I thank him for sharing the fruits of his labour.' **Richard Bell, artist and activist**

'Our Ancestors cry with joy.' **Maurial Spearim, Gamilaraay actor and playwright**

'Sharing your understanding is creating a space for others to be "present". It's not easy to say out loud what others may not understand or might not want to understand, and it takes courage to say out loud what others might whisper. We are always learning, thank you for sharing.' **Aunty Cilla Strasek, Gamilaraay-Yuwaalaraay Language educator**

'A beautifully written and valuable work of Gamilaraay knowledge, which brings the rich nuance of Indigenous philosophies to the fore.' **Aurora Milroy, Palyku legal scholar and Director of Policy at the National Native Title Council**

THE EAGLE & THE CROW

JM FIELD

First published 2025 by University of Queensland Press
PO Box 6042, St Lucia, Queensland 4067 Australia

University of Queensland Press (UQP) acknowledges the Traditional Owners and their custodianship of the lands on which UQP operates. We pay our respects to their Ancestors and their descendants, who continue cultural and spiritual connections to Country. We recognise their valuable contributions to Australian and global society.

uqp.com.au
reception@uqp.com.au

Cover design by Jenna Lee
Typeset in 12.5/17.5 pt Bembo Std by Post Pre-press Group, Brisbane
Printed in Australia by McPherson's Printing Group

University of Queensland Press is assisted by the Queensland Government through Arts Queensland.

University of Queensland Press is assisted by the Australian Government through Creative Australia, its principal arts investment and advisory body.

A catalogue record for this book is available from the National Library of Australia.

ISBN 978 0 7022 6912 7 (pbk)
ISBN 978 0 7022 7075 8 (epdf)
ISBN 978 0 7022 7076 5 (epub)

University of Queensland Press uses papers that are natural, renewable and recyclable products made from wood grown in well-managed forests and other controlled sources. The logging and manufacturing processes conform to the environmental regulations of the country of origin.

Contents

Acknowledgements

THE RESEARCH AND WRITING OF this book was done over five slow years. The number of people, then, whose presence I feel in it are too many to name.

Some particulars, however, must be: my sincere thanks for being part of the *Kin Review Process* to Joshua Waters, Aurora Milroy, Maurial Spearim, Warraba Weatherall, Aunty Cilla Strasek, Aunty Amy Creighton and my mum, Aunty Pat Field.

The greatest thanks, however, goes to my eldest sister, Jayde, without whom *The Eagle and the Crow* would not exist. She read almost every version and continued to ask – at the right time – the right questions.

My gratitude to the above people should not, in any way, be misunderstood: any fault with the book rests solely with me.

Giirr maaru Jared, bamba ngaya nginunha
winangaldanha. Burrulaa nhama garay
guwaaldanha, waalbala Dhawungu
nhama garay guwaaldanha.
— Aunty Cilla Strasek

DHULANBAA I

1.
This will be difficult to express in the language of my coloniser. I will fail.

2.
I will try.

3.
Dhulanbaa is the time of the Black wattle.
Dhulanbaa is the place of the Black wattle.

4.
There is no time without the place, and no place without the time.

5.
We do not mark each beat and swallow it whole.
We are rhythm people.

6.
When you die, said Garruu, you may be reborn not as person or animal. Not even as tree.

You may, instead, be reborn as place.

7.
This makes mining murder.

8.
This also makes us strong.

9.
When I say we are rhythm people, I do not mean that we just track them. And certainly not force them.

What I mean is that we *participate* in a rhythm that is larger than ourselves.

10.
Those who do not may perturb it, *yawu*. But they cannot break it.

Gamil.

11.

Take solace, my people. Rest easy.

12.

Wake early, my people. There is much to do.

13.

This will be difficult to express in the language of my coloniser. I will fail.

14.

I will try.

1

Gamilaraay locatedness and the limits of Tindale's map

IN ORDER TO TALK ABOUT Gamilaraay kinship, I first, of course, need to talk about who we the Gamilaraay are. To do this, I need to *locate* us. The first impulse, for the non-Gamilaraay reader, is to vaguely point to Tindale's map, to find the blob that Tindale himself said that we were, and perhaps situate that within well-known colonial borders: somewhere in the south-west of Queensland and the north-west of New South Wales. This is wrong, as we will see, but in the way that mathematicians appreciate wrongness: it is usefully wrong. If we understand maps to be models of the world, we must acknowledge that they are approximations. Approximation, in turn, if done in a clever way can be handy indeed: a butterknife is not a screwdriver, but in a pinch it is, approximately.

Having laboured to talk about things that are *usefully wrong*, I'm now sorry to complicate things further by pointing out that there are few concepts more subjective than that of *usefulness*. Useful for what purpose? Useful to whom? Useful when? For us Gamilaraay, Tindale's map has very little utility; it has helped, if I'm generous, in convincing some settlers that we are not in fact Aboriginal. By this I mean, of course, that there are no Aboriginals at all, that instead we are a continent of culturally, physically and historically diverse peoples. Otherwise put, the convenient fiction of the Aboriginal is an *approximation* that is *useful* but rarely to us.

Indeed, the category originally borrowed from botany and taxonomy made it possible to include us, in one easy sweep, into colonial legal systems. In fact, much to our detriment, it first formally entered legislation alongside blood quotum classifications in the colony of New South Wales in 1839, followed by the colony of Victoria in 1864 [6]. It wasn't until the late 1950s, sustained by the growing science of eugenics, that states ceased regularly legislating all forms of inclusion and exclusion (to and from benefits, rights, places etc.) by reference to proportions of

Aboriginal blood [6]. My mother, for instance, then considered full-blood, was not permitted to freely leave the mission for much of her youth.

Now, that last example I stretched out some way, but the aim was to drill home just how dramatically approximations can help or harm. Tindale's map, I said, was useful to approximately *locate* us, but the very many Blackfulla groups that have been erased entirely from iterations of it – such as the Yuwaalaraay – might disagree. Their stories, culture and customs were known to settlers well before Tindale (see, for example, the work of Langloh Parker [12, 13]), so their exclusion from any version seems sloppy indeed. His map and later iterations, however, are disagreeable in more fundamental ways, too: us Gamilaraay do not wayfind with lines on a page. Our maps, and therefore our models of the world, are instead stories and songs. Our *external* borders, then, cannot be as strict as a fence. Songs are coarse-grained and encourage a degree of fuzziness. In turn, where we are, which in part determines who we are and are not, is also fuzzy. The western Gamilaraay might consider *some* of the Yuwaalaraay, who are made of many of the same stories and songs, to be in many ways Gamilaraay. This view need not be held by more eastern Gamilaraay. This is all to say that while Tindale's map attempts to say who is Gamilaraay, who is Wiradjuri, and so on, our worldview more accurately determines who *isn't*. Those who aren't Gamilaraay are those who *all* Gamilaraay agree are not Gamilaraay; they are those overwhelmingly composed of different stories. This says nothing, mind you, about the finer-grained problem of whether or not an individual belongs to a particular family group. A great many people now claim the former without first being claimed by the latter, a particular problem so knotted that it neither can nor should be dealt with here.

Now, the keen reader might note that I continue to say *locate* and that our *where* only partly determines the *who*. The reason for this, which is another flaw with Tindale's mapping, is that Gamilaraay conceptions of place are largely not detached from time. Gamilaraay academic Joshua Waters explains this best:

> For example, *walaaybaa* can mean 'place where we camp(ed) or will camp in the future', and simultaneously 'the time when we camp(ed) and/or will camp in the future'. Similarly, *baan baa* can be 'place of the mistletoe', or simultaneously 'time

of the mistletoe' because the place where it grows also implies the time it is growing or existing, in a certain place at a certain time [23].

Gamilaraay people do not *locate* in time nor in space but instead both at once, through events and their cycles.[1] In contrast, Tindale's map is a relic frozen in the past; a fact that is no doubt useful to the colony whose existence is predicated on our displacement. Purposes aside, this map – so disconnected from time – is *incapable* of locating us as we do ourselves.

The map of Tindale, then, is neither screwdriver nor butterknife. Does that mean we ought to discard it entirely? Certainly not. Despite the many ethical question marks over the original work that led to the map (see, for instance, even the first two pages [20]), it allows people like me to crudely and lazily point out our whereabouts for the non-Gamilaraay. This is fine, as with all models, so long as we articulate its limits. Tindale, in fact, deserves some grace, for without a notion of Gamilu wiibidi his mapping was doomed

1 I am not suggesting that we do not have tenses; we certainly do and they are extremely interesting in the way that they do not map to English ones. I am saying that responses to 'Dhalaanda?' or 'Where are you?', such as 'Gundhidha ngaya' or 'I am at home', traditionally encode time and place at once. Gundhi, our word for 'home', is also our word for 'stringybark': you cannot make gundhi if you are in time and space distant from a grove of these trees.

to find us. Gamilu wiibidi – where all Gamilaraay originate from, will go and in part remain – is our notion of time outside of time and place outside of place. For the physicists reading, I note a more direct translation of this compound noun: before, outside of or in absence of light.

Without a more solid foundation of Gamilaraay philosophy, the above characterisation of our location might understandably be a bit disappointing. I will offer, instead, a more political approximation. To do so, I note, as Gamilaraay poet Lorna Munro did in conversation with Nayuka Gorrie for the Sydney Writers' Festival in 2021, that our word yilaalu can be translated as 'once upon a time in the past and the future' [11]. The essence of this word is that it is a distant time-place location. I also note that it is no accident that gamil, our word for 'no', is included in the name we give ourselves. Gamilaraay people, then, are located and defined by the lands where now and yilaalu, gamil is heard, understood and final. It is where we are sovereign.

Note on orthography

Much to the frustration of others, 'Gamilaraay' has been be spelt some two hundred different ways, all of which are entirely valid. One of the earliest, for instance, by missionary William Ridley was 'Kámilarói' [16], with the accent on the second letter indicating it isn't quite the English 'a'. It is, instead, somewhere between 'a' and 'o'. The accent on the second last letter serves a similar function and separates the final 'i' vowel sound. Another common source of variance is the split between 'G' and 'K' as seen in 'Gamilaroi' and 'Kamilaroi'. Again, this is because the beginning sound is neither the English 'K' nor 'G' but rather something in-between.

While the initial spellings were all created by our colonisers, the very many iterations that have evolved since reflect instead the important work of reclamation. It is in this sense that all versions are equally acceptable. Indeed, the multitude of spellings mirror not only our diversity (our homelands are, in fact, more than twice the size of Belgium) but also a refusal to be homogenised. For those new to us, however, I understand how this might cause some difficulties. For this reason, I list below some of the more common orthographies: 'Gamilaraay', 'Kamilaroi', 'Kamilarai', 'Gamilaroi', 'Gomeroi', 'Gumilaroy', 'Goomeroi'.

Note that the spelling employed throughout this book – 'Gamilaraay' – is different to the others insofar as it is not intended to be read in English; just as French uses the same letters as in English but with different sounds associated with them, the same is true here. The 'aay' combination, for instance, in *standardised Gamilaraay* makes the sound of 'i' (the name of the letter) in English. The first 'a', because of the surrounding letters, is much closer to an 'o' in English. To *hear* how I intend 'Gamilaraay' to be pronounced see sentences 941 and 942 in the online dictionary,[2] where Uncle Arthur Dodd and Uncle Fred Reece can be heard pronouncing it. This is almost the same pronunciation procured from my grandfather, Uncle David Spearim, who was recorded speaking with Uncle Bert Draper. This, as with many other recordings, can be listened to in the AIATSIS archive.

While I think there is value in standardising the written language as a whole, I don't believe this to be true of the very names we give ourselves. The above is not then, I must reiterate, an invitation to police others on spelling or pronunciation.

2 Note that all mentions to 'the dictionary' refer to what is currently known as *Gaman guladha Gamilaraay, Yuwaalaraay, Yuwaalayaay*, produced and owned by John Giacon and David Nathan. Both this name and its web address have changed many times and I'm told will change again soon.

DHULANBAA II

1.
This will be difficult to express in the language of my coloniser. I will fail.

2.
I will try.

3.
Gamilaraay language, it can be argued, has no simple future tense.

4.
Instead, we speak of uncertainties:
It might rain.
She might die.
She might live.

5.
In this sense, it is a language interested in truth.

(We also, unlike English, have no passive voice: *He killed her* cannot become *She was killed.*)

6.
A provocation: how then, if we are honest, can the wattle be late?

7.
I am not trying to suggest that lateness does not exist. I am saying instead that it has little to do with the wattle.

8.
Lateness, to be painfully clear, belongs to the person who demands one present and receives another.

9.
A consequence: the wattle is not late, has never been late.

10.
The wattle is located.

Dhulanbaa.

11.
We are located.

12.
Do you understand the seriousness of this? It is why the old Aunties cannot be rushed, will not rush others.

13.
It is why we must not rush others, must not be rushed.

2

Gamil relationships to knowledge

KNOWLEDGE IS NOT A RIGHT. I mean this in the sense that non-Gamil people are not owed open access to our ways and knowing, and so for this reason I have intentionally left a great many details out. But I mean it also in the sense that other Gamilaraay mari are not owed any specific learning: the details – if they are meant for you – must be got through your old people. For millennia, ceremony – both big and small – ensured that knowledge found us when we were responsible enough to have it, and this has not changed. Though it sometimes looks very different, ceremony, like us, is still very much here. It remains in the big corroborees to sing the rain, as well as the seemingly modern moments of making tea for your

Aunties. In many ways, to Gamil people who crave the details, I am simply saying this: make more tea for your Aunties.

I am also saying that we must be wary of what we share, and with whom. There are a great many people who traffic in our knowledges: for some, their *credentialised* expertise is simply a way to pay the mortgage; for others, it is born-again mission managerialism. Most often, in my experience, it is a sick mixture of both. Gamilaraay artist, activist and essayist Richard Bell hinted at this in his powerhouse essay, *Reductio ad Infinitum*, when he pointed out:

> the clear capitalist tribal order that ranks white specialists as more knowledgeable on Aboriginal art and identity than Aboriginal people themselves [2].

Universities, of course, are almost solely responsible for this truly ugly practice.[3] Indeed, credentialism is, in my view, the very highest form of gatekeeping precisely because the knowledges *purchased* take work to understand, and so in some sense seems noble,

3 I say 'of course' because the foundational function of universities in settler-colonies was to spread imported knowledge on stolen lands, thereby supplanting that which already exists. It might well be argued this remains a working definition.

but also because who gets access is limited across the invisible lines of class. The result is that those who are supposedly experts of us feel just and right in their place in the hierarchy, while both now and historically we are quietly and generationally mostly excluded from even joining the ladder.

While credentialism may make a weapon of all knowledges, in the case of ours, it is an especially tired tragicomedy. The comedic element comes from the fact that what is being bought is rarely truly our traditional knowledges. Instead, they are curated tidbits or things preserved apparently for our benefit. Of course anything curated is incomplete and anything preserved is presumed dying or dead – of which we are neither. This is all to say that these things ripped from context and siloed into distinct disciplines are but a pitiful caricature of our knowledge networks. The tragic part comes from the fact that white expertise has been, and continues to be, used in nefarious ways against us. The most obvious, and in my opinion most outrageous case, Bell wastes no time to highlight in the very same essay:

> Aboriginal people can't turn up to a land court and have our rightful claims heard without the

verification of some white scholar from Sydney, New York or Melbourne. That is the reason anthropologists are still on our land. The onus should always have been on white title holders to argue for their occupation of our land under claim [2].

A less obvious, but increasingly worrying, instance can be found in the Gamilaraay language revivalist movement. Spearheaded by non-Indigenous linguists, a whole host of so-called neologisms are appearing. I say so-called because neologisms – or new words, meanings or expressions – appear organically while what is occurring here is instead forced and external. A concrete example is the translation of our word 'gilay', meaning 'moon', to also mean 'month'. This seems rather natural and intuitive, and indeed it is, but it is also incredibly dangerous. We did not have a word for 'month' because that is not how we viewed the world; overwriting gilay in this way is nothing short of forcing settler-colonial understandings of time on to and into our language. In short, it is rendering their way of measuring and marking time as the standard at the very real risk of losing our own.[4]

It is also deeply lazy. When a Gamilaraay child asks what our word for 'month' is, the first and easy option is to say that it is gilay, precisely and *only* because a white expert said so. The alternative is to be generous with these little ones, and provide nuance. Indeed, why should they ask about the warm wind, Yarragaa, that kisses some of the trees to make them bloom if instead a direct translation to the European idea of spring is given? An idea, mind you, that due to our longitude and latitude makes not a lick of sense on Gamilaraay Country. Spring, however, is not the only offender: whitefulla time more generally – *so displaced from place* – is a bizarre thing indeed on most of this continent. Clocks and calendars on walls now tell the trees they're late with barely a thought given to the absurdity involved in such an imposition.

Now, an argument I've heard for the creation of these very many new words is that so few of our

4 The devil's advocates reading (which devil, I wonder) might well argue that we did have a conception of 'month'. This is true, *approximately*. In fact, our version very beautifully mapped phases of the moon to our bodies. However, the current Western conception of a 'month' is economic, and therefore divorced from natural phenomena. If I dump you on my country with just a calendar, it will be very hard to tell when the new day starts (arbitrarily, for whitefullas, at so-called 'midnight'), let alone predict and organise around what the moon will do.

original ones survived our colonisation. In turn, for a language to be functional day-to-day a threshold number of words are necessary. This is sound reasoning. The issue arises with what is then done with this thinking: namely, certain non-Gamilaraay people have taken it upon themselves to set the timescale of our language revitalisation. This benevolence, however, is the one and the same that led to our language being murdered by Christian missionaries in the first place. The difference is that the urgency to save our souls – while taking our land – has been replaced with the urgency to save our language but not, it seems, our culture. This is all to say that the benevolence in question, as it so often does, remains assimilatory in nature [25].

While the occupation of our land appears *fait accompli*, I hope it is now clear that the war against and weaponisation of our knowledges actively trundles on. This is, of course, no accident. The misuse, abuse and disruption of Gamilaraay knowing is the means by which land may become *unstolen*. Otherwise put, the land cannot be meaningfully returned if we become culturally, economically and philosophically the same. That is the true function of assimilationism. The irony of the above mentioned weaponisation is that it speaks directly to the rather loose grip others have on our lands: for why else would one attempt to render land *unreturnable* if you were so certain of your hold?

Clearly, there is great potential in the knowing of ourselves. I do not at all mean this, however, in the Western hyper-individualised sense: there is no Gamilaraay knowing, or indeed knowledge, without others. It is properly *expressed* collectively. Put another way, we must continue to make tea for our Aunties: our very land depends upon it.

Note on publishing

On the one hand, the written word and machinery of publishing is very useful to us Gamilaraay people. It allows our thoughts and intentions to not just be presented to our generation but to those who may come after. On the other hand, publishing is by nature a public endeavour: people who are not Gamilaraay will read this work. Many things about us, however, should not be known by others. For this reason, we must aim to write in a way that rewards the Gamilaraay reader but is, in some sense, obscure to others. This is very tricky indeed, but it is not new to us: we developed – long before our colonisation – an entirely separate language that only initiated individuals could understand. The Gamilaraay person, by using English with cunning, honours this tradition. This is all to say that throughout this book holes will be left *on purpose* in an attempt to remain a good ancestor. It is up to the Gamilaraay reader to sit with the correct person to complete that knowledge. It is only in that way, I am certain, that we can keep our knowledges safe.

DHULANBAA III

1.
This will be difficult to express in the language of my coloniser. I will fail.

2.
I will try.

3.
My hand, my shadow, my thigh, my meat: in Gamilaraay, we do not use possessive pronouns – words like 'my' – for things inseparable from ourselves.

Mara, malawil, dharra, dhii.

4.
'My' and 'your' tend to be found, instead, when there is responsibility involved: *nginu birralii, nginu bilaarr, nginu gunii, nginu guuyay.*

Your child, your spear, your mother, your mood.

5.

This last one I especially love: *yaama nginu gaba guuyay?*

Do you have *your* good mood?

6.

Otherwise put, forget: how are you going? Forget, too: are you in a good mood? We recognise, instead, that there is responsibility involved.

My foul mood is mine alone.

7.

I am not trying to suggest that we ignore the conditions we find ourselves in: we did not ask to be colonised.

I am suggesting that we can and must be intentional with what comes next.

8.
This should not at all be read as cheerful moods ought to always be picked. Growly Aunties teach many things.

9.
What I am saying is this: refuse, like your old people, to be a leaf in the wind.

We are instead the trees.

10.
Bloom when you can, seed as you must.

3

Kinship structure and robustness to colonialism

IT IS DIFFICULT TO START a chapter on the specifics of Gamilaraay kinship because the topic is both vast and complex; for instance, properly understood it includes non-human kin. Mostly, however, it is difficult because linear explanation is not really our way – as noted in Chapter 1, we anchor ourselves in the world with events and cycles. We are also, from a young age, encouraged to be systems thinkers; which is to say our search is for unexpected links instead of organised silos. A unique start and a natural finish, then, is neither necessary nor particularly sensical. Otherwise put, the web catches the fly, not the thread. Nonetheless, the written word has forced my fingers: I begin, arbitrarily, with ngali – 'we two'.

Now, ngali is a word that English speakers might find odd. Unlike ngiyani, which is closer to the English 'we', ngali is a pronoun that restricts to dual pairs. For instance, ngali guwaymadhan would be translated as 'we [the two of us] are dark blood', but not 'we [more than two, the group] are dark blood'. This highlights the significance of these sorts of relationships in our social organisation: rights, restrictions and – more importantly – responsibilities occur not just at the level of the individual, but also as units consisting of pairs. Again, Gamilaraay academic Joshua Waters explains this best:

> In a K/Gamilaroi governance structure, fractals inform the scaling of relationships from ngali (two people as a relational/kinship pair) all the way out to ngiyaninya (us all, as a network of networks). This is explained as, while every person is in full control and possession of their own autonomy, they are bound by the protocols and requirements of each of the individual ngali (i.e. the two of us) relationships which they have with the other people in their immediate kinship network. [24]

These restriction/responsibility networks do not stop here but, as Waters explains, scale outwards from ngali pairs to family networks, to clans and to nation groups. In turn, nation group pairs are also bound by protocols and requirements between each other just as all pairs at lower scales.[5]

The focus, however, of this chapter is a much finer scale. To pull us back, I return to the *two of us* and note that ngali relates to our word 'ngalirr', meaning 'umbilical cord'. In turn, this derives from Ngalirr, one of the powerful wives of Buwadjarr, the All-Father. It is said that she gave us – fashioned from her own body – the first umbilical cords, and so after our creation by Buwadjarr, enabled our continuation. Ngali, then – and the relationships within and between such pairs – emphasises the connection between them instead of the individuals themselves.

This connection-based thinking is really the key to understanding our kinship. For instance, while in English 'Aunty' may mean either your 'father's sister' or your 'mother's sister', in Gamilaraay there is a separate word for each. The same is true for

5 Note, however, that I'm strongly of the *opinion* that the notion of a nation (i.e. organisation at such a high scale) is quite new to us. Nonetheless, it is a necessary and clever *response* to our colonisation: if we wish to be heard by the colonial state (a debatable idea in itself) we must speak from lofty heights.

'niece', 'nephew', 'son' and 'daughter', which change not because of who they are but how they connect to you – either through sister or brother or self. The word, for example, used by *both* my sister and me to traditionally indicate my children is not the word ngali, the two us, would use to indicate her children. Otherwise put, niece, nephew, son and daughter are not natural distinctions that exist in old Gamilaraay.

Indeed, when my mother's sisters call me 'my son' in English, this is not exaggeration but instead a maintenance, in part, of this kinship relation. There is, unfortunately, a great flattening of our languages to fit colonial worldviews in this regard, as touched on in Chapter 2. Another example, which settler linguists have a lot to answer for, is the translation of 'baawaa' to simply mean 'sister', when strictly speaking it is 'senior sister'. These may seem finicky things to take issue with, but truly it is a mistake to view this flattening as anything other than the continuation of the colonial project. Indeed, what good is a renewed Gamilaraay language if we use it to think and relate like our colonisers do? As Oodgeroo Noonuccal much more elegantly put it, as far back as 1966:

Pour your pitcher of wine in the wide river
And where is your wine? There is only river. [22]

Again, it seems I have strayed off the narrative path. To assure you that the web is indeed being built, I return again to ngali. This time, however, I come to the first and most literal such pairing: that between birth gunii, or mother, and child. It is this connection that determines all other attributes, and therefore connections, that Gamilaraay people have whether they know it or not: moiety, blood type and meat.

Now, some of these terms some readers may be familiar with, others perhaps not. I will assume nothing, and instead ask for patience. Moiety – the highest level of division – arises from a splitting of the world entirely in two: people, animals, trees, waters and even winds. All things are either Wudhurruu or Yanguru. For our purposes, all Yanguru are Guwaymadhan, and all Wudhurruu are Guwaygaliyarr – this is what is meant by blood type. The direct translations of Guwaymadhaan and Guwaygaliyarr are in fact dark or heavy blood, and light blood, respectively. The moiety names proper, much to the frustration of a younger me, do not have translations. They do, however, have representatives:

Yanguru are represented by Kaputhin the wedge-tailed eagle, and Wudhurruu by Dilby the crow, at least amongst the northern Gamilaraay. Again, these words do not have translations, nor will you find them in the Gamilaraay dictionary: they do not mean all crows nor all eagles. Instead, they are a particular ancestral crow and a particular ancestral eagle.[6]

Finally, there is dhii, or meat, which many other mob tend to call 'skin' and which sit within each blood group. This last one, however, is tricky in that it also means a particular totem that is inherited based on the matriline you belong to.

First, let's focus on the meaning that is closer to 'skin', which, amongst other things, determines who you can and cannot marry. In the Gamilaraay system, there are four distinct dhii groups inherited through matrilines, with the initial division being by blood type: guwaymadhan mothers produce guwaymadhan children, and guwaygaliyarr mothers produce guwaygaliyarr children. The blood types are further

6 This view of the relationship between Yanguru:Wudhurruu and Kaputhin:Dilby, given to me by my family, I suspect to be a teaching tool. Older accounts seem to take Kaputhin and Dilby not as emblems, as such, but as alternative words to the moieties themselves. The Yanguru:Wudhurruu pair, I suspect, was used to interact with our neighbours to the north (where these terms are also found), whereas Kaputhin:Dilby were employed to interact with those to the west and south, many of whom have an Eagle:Crow moiety split. For further reading on this see Chapter 17 of [21], and citations therein.

split into two, to give the total four groups: yibadhaa[7] and buudhaa, amongst the guwaymadhan; and maadhaa and gabudhaa, amongst the guwaygaliyarr. The masculine versions of these terms are yibaay, gambuu, marrii[8] and gabii, respectively.

The particular subtype or dhii you take is also determined by your mother, but in a slightly different way: you maintain her blood type, but are born into the opposite subtype. Put more concretely, yibaay ngaya or I am yibaay, because my mother is buudhaa, and in turn because her mother was yibadhaa. This is summarised in Figure 1, with Gamilaraay names removed, in order to eventually ease mathematical thinking. I will touch more on this later, but it is worth noting that in Gamilaraay kinship there are cycles of dhii matrilineally amongst blood types. Save a few rare exceptions, this cyclical nature of kinship is something especially unique to our continent, which

---

7 The letter 'y' before the letter 'i' in standardised Gamilaraay is *almost* silent. It is a little closer to a soft 'h', and may change depending on the end of the previous word. This will be no surprise to Gamilaraay people whose mothers pronounce 'eggs' as 'heggs' and 'house' as 'ouse'. Though the mission managers banned our language, many elements from it were very cleverly snuck directly into our variety of English.

8 Some Gamilaraay people believe that this is the source of the word 'Murri' by way of another Gamilaraay word 'mari' meaning 'person'. I fall into the camp that suspects this to be false: if you listen, for instance, to recordings of the old people saying the dhii name, marrii, the 'r' is rolled and the vowel sound extended. This is not true when they say 'mari'. The direct connection between 'Murri' and 'mari' is a little more convincing.

has remarkable implications for our genetic structure. For older anthropological accounts of the Gamilaraay system see [5, 12, 15],[9] whereas for modern primary source treatments see [17, 18]. Recordings of Uncle Fred Reece (born in the 1890s, and who was recorded in the 1970s) pronouncing the dhii names can be found under sentences 378 and 379 on the online dictionary.[10] I strongly suggest you listen to these before continuing, in order to start with the correct pronunciation.

Now, strictly speaking, I am not simply yibaay but instead yibaay dhinawan ngaya. This is where the other meaning of dhii springs from: I am emu yibaay. This level of distinction arises, of course, because Gamilaraay lands are large; the yibaay ngurray or black snake yibaay are different to my emu yibaay. This level of dhii also comes through matrilines, but indicates a finer scale family connection. It also, I suspect, is where the overlapping meanings of dhii come from: giir ngaya yibaay dhinawan is the more

9 Note, however, that one of the many reasons I wrote this book was so that future Gamilaraay people will not have to engage with these anthropological accounts – they are often violent, sometimes wrong and almost always unethically sourced.

10 These can be accessed either through the main sentences page of the online dictionary, or by searching any of the terms via its main search function. To get the audio file, you then need to click on the little text bubble at the end of the dictionary entry.

complete answer to the question: minya nginda dhii? Or: what meat are you?

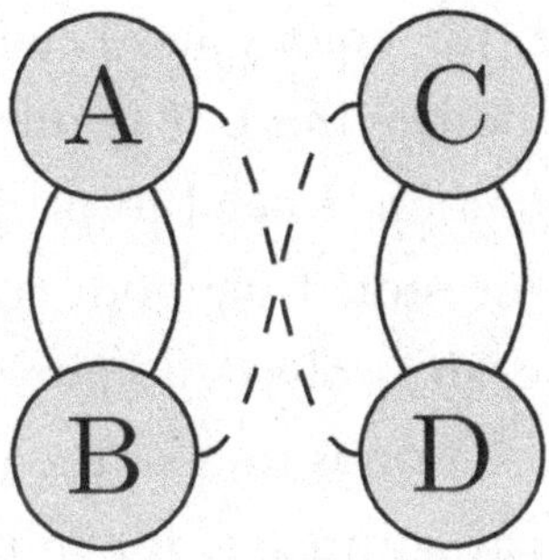

Figure 1: Kinship dynamics of the Gamilaraay system for *first preference marriage*. Solid lines represent matrilineal inheritance of dhii whereas dashed lines link dhii groups who are permitted to marry. Individuals are only permitted to marry someone of the opposite blood type (CD type people, if an individual is of type AB, for example) and opposite subgrouping to their father (C type people, if an individual is B type). Note that this ensures that an individual can only marry someone who is of a type different to themselves and both their parents. By virtue of inheritance rules, it also ensures individuals do not marry siblings, aunties, uncles and maternal cousins of any order related through women. Note that second preference marriage will be dealt with later.

At this stage, I am sure that there are a great many Gamilaraay people who, tracing their Gamilaraay essence through fathers, might begin to feel uneasy. And, of course, the colony has been most cruel in its attempts at destroying our connections to each other. For this reason, I need to pause to settle any concerns: our ancestors have both encountered this problem before and solved it. For people in this situation, one solution is to take their dhii as if they had a Gamilaraay gunii (so, for instance, opposite blood type of their father, a new matriline totem and so on) via adoption. They must, however, not discard their father's matriline totem, which is the second half of dhii, for reasons that will become clear in the next chapter.

Now, of course, there are many ways one might imagine this problem to be solved.[11] This particular solution, however, achieves three things: first, it maintains connection through meat (dhinawan, yurrandaali, and so on). Second, it protects the overall structure of the system itself: matrilines continue to be followed, instead of patrilines. Third, the

11 Indeed, it *has* been solved in many ways, both on our lands and on other parts of the continent with varying degrees of grace. Do not read the above solution as the authoritative one, which in fact does not exist.

adoption process shores up community connections: gaining another mother also means gaining more siblings, aunties, cousins and so forth.[12] None of this is to suggest, however, in any way that marriage wrong-way could or should be worked around. Indeed, the punishment traditionally for such behaviour was extremely severe – banishment at best. This is all to say that the solving of what I call the bubaa guway or 'father-blood' situation is a grace not to be taken lightly. Indeed, in the regrowth of our kinship system it is worth thinking very thoroughly about when we ought to be generous, and when we should not.

Now, when I speak of regrowth, I mean it in the sense that whether a Gamilaraay person knows it or not they have a place in this system. If it has been forgotten in their family – and there are a number of reasons that may occur – only two things are needed to remember: a 'seed', and information of how they relate to that seed. The latter is knowledge that has been carefully guarded on the missions; Aunties

12 To be clear, traditional adoption practices do not apply to Gamilaraay people who have a Gamilaraay father and, say, a Muruwari mother. The reason is that all of our neighbours, and many beyond that, also have a four-group kinship system that allows and historically encouraged intermarriage. The northern Gamilaraay, for instance, where my grandmother is from, have very old ties with the Bigambul to the north.

there contain libraries in their heads of who married whom, when and where going back generations. In some cases, those particular mental records stretch back to pre-invasion times. 'Seeds', on the other hand, are individuals from families who secretly, and at great risk, passed on kinship knowledge. In doing so, they kept not only their knowing of where they fit, but also the potential of others to remember. The gift that this regrowth potential is truly cannot be overstated: our kinship system, by design, is robust to disturbances. Indeed, the architects of it, our old people, created a way of relating, and therefore organising, that colonisation could not break.

Before I give a concrete example of regrowth, to help those who need it, I must first pause here to remain a good ancestor and be unequivocal: this regrowth work cannot be done simply with colonial records. Otherwise put: *you cannot be in kinship alone.* I worry, worse still, while writing that those who have no interest in good relations, but simply wish to bolster their legitimacy, may abuse my work. This, however, I cannot guard against by myself, but must and will rely on the collective knowing of other Gamilaraay people. Put another way, in the mammoth task of nation (re)building, we are forced to bet any benefits against such costs or simply do nothing at all. The bet, however, I'm certain is skewed in our favour: giirr ngiyani walanbaa.

All that said, I can now in good faith get on. Suppose, for instance, that you have spoken to an older Aunty, and she has told you that your father's mother's mother shared a mother with my great-grandmother. In this case, I am thinking of my mother's mother's mother. This may sound like an odd example but, I assure incidental non-Gamil readers that in our homes, these seemingly distant connections are both well remembered and honoured.

Now, because my meat is yibaay, from Figure 1, I know that my great-grandmother must have been buudhaa. In general terms, suppose that I am type A, my mother must be type B, her mother must be type A, and her mother type B, following matrilines. Of course, the causal arrow goes in the opposite direction – I am yibaay because she was buudhaa, not the other way around, but this demonstrates a sort of knowing we have of our matriarchs long gone that is rare in other places.

But back to our grannies: having established that mine was buudhaa, it tells me that her sister – your father's mother's mother – was also buudhaa. As above, following matrilines in Figure 1, we are then certain that your father, even if he does not know it, must be gambuu; this makes him either like an uncle or a nephew to me. Now, because he is gambuu, if we assume marriage proper-ways, your mother is necessarily maadhaa or gabudhaa. Following her matriline to you, you are then firmly located to be either gabudhaa or gabii if she is maadhaa, or maadhaa or marrii if she is instead gabudhaa. You are now also, more importantly than anything written so far, *a new seed.*

The significance of this cannot be overstated: one seed creates not one but many more in turn. The regrowth rate of our kinship pattern, once woken up, can outpace centuries of colonisation in a very short time. Better still, the method to rebuild our kinship system is contained in the very knowing of its structure; a structure, mind you, that is packed full of useful symmetries. For instance, understanding how to propagate kinship forward through children instructs us equally on how to propagate backwards to gather distant relatives who might have forgotten their location. This regenerative property of our kinship is well worth pause for reflection.

For example: how many times before have we rebuilt our kinship? How many times more will we rebuild it again? Behind these provocations is something sad indeed; imposition, invasion, destruction of culture, and so on. Despite this, these questions are, to my mind, cause for great comfort; they speak to the sheer scale of Gamilaraay *presence* on Gamilaraay lands. In turn, they shrink the size of the seemingly gargantuan task at hand. Otherwise put, it is very likely that we are neither the first nor the last Gamilaraay to regrow the pattern: we simply need to persist.

A few notes on marriage

When I say marriage I do not mean it in the Western Christian sense. Traditionally, for instance, Gamilaraay marriage did not bind us to other people with quite the same strictness – marriage, usually, simply meant that you shared a fire with someone in a certain way. To the horror of many early missionaries, this was often done not just with one person. The former is reflected, to this day, in many Aboriginal Englishes where 'married up' tends to mean that two people are serious to the point of living together. The latter – i.e. accepted non-monogamy – is all but gone on our lands. None of this is to suggest, however, that we did not bind people in our own ways; my grandparents, for example, were promised to each other along very old political lines before they were even born.

Another important thing to note is the practice of second preference marriage [15]. This arose, and was occasionally permitted, because of the additional marriage tendency of avoiding the same meat (dhinawan, ngurray etc.) as your father. For example, yibaay should be wary of marrying gabudhaa, even though they are marriageable 'skins', if the one who is gabudhaa and the father of the one who is yibaay are both owls.[13]

In turn, this means that a reluctant second preference marriage may occur with someone who is in the same 'skin' grouping as your father (so, for instance, marrii), but different second meat (an emu instead of an owl, say). This is only possible, however, if you have not already had a first preference marriage. Conversely, if a second preference marriage occurs first, you cannot then have a first preference one. The reason in both cases is that the one or the

13 This will be elaborated on later, but the reason is that the second part of your meat tells a more immediate family history.

other grouping then *becomes* a nganbiir or crossways relationship. In some instances, this is known as mother-in-law avoidance. The above highlights another important quality of our kinship: it is not static, but instead entire groups of people are in a state of flux as we and others progress *through* relationships.

DHULANBAA IV

1.
In the Gamilaraay-Yuwaalaraay dictionary, a *gunj* to be sure, *gadjigadji* is listed as regrowth.

It is the many small trees that push up after a flood.

2.
'*This word indicates that there may be an unrecorded word, gadji*,' the dictionary says.

This is an exceedingly reckless way to talk about colonialism.

3.
Make no mistake: words were not simply *unrecorded* and rarely did they die. They were instead murdered.

4.
No matter, we have gadjigadji.

5.
By which I mean, there is good, strong regrowth.

6.
Still, soon it will need to be burnt back.

7.
This will be difficult to express in the language of my coloniser. I will fail.

8.
I will try.

4

The form and function of Gamilaraay totems

REGRETTABLY, THERE IS GREAT CONFUSION, even amongst ourselves, surrounding our totemic system. Part of this confusion arises from a misunderstanding of form: we have different *types* of totems for different purposes. Overwhelmingly, however, the issue arises from the function of one of those totems, the second half of dhii, sometimes called mara (meaning 'hand'),[14] and its relationship to marriage. This is, without a doubt, due to certain older anthropological texts that really ought to be retired. In lieu of concretely stating what is right and

14 Note that Austin [1] referred to this type of matrilineal totem as 'Yaruthaga', now standardised in the online dictionary as 'Yarudhagaa'. I opt, however, for what I was taught by my old people.

what is wrong, which is not the Gamilaraay way, I will instead here *show* what can be learnt from the knowing of mara. This way, other Gamilaraay mari may either incorporate what has been given or, without bitterness, discard it entirely.

To begin, recall that the second half of dhii – here called mara – is inherited matrilineally: if your mother is dhinawan then you are dhinawan. Of course, the first half of dhii – yibaay, for instance – is also inherited matrilineally but in a slightly different way: you take the opposite dhii in the *same* blood group. This means, in turn, that blood group is inherited as directly as mara. The difference is that while there are only two blood groups, Guwaymadhan and Guwaygaliyarr, there are very many hands.

For this reason, mara is much finer scale and can be used to track relations over the vast distances that result from unions: traditionally, sons stay on their father's land whereas daughters, upon marriage, migrate. These distances should not be underestimated: while your mother has come from only one other land, her mother came from a different one still, and her mother before her. These moves accumulate very quickly, and not only have descendants become separated by large physical distances but also metaphorical ones.

The daily particulars of who married whom, for instance, are hard to keep up with when this physical separation combines with generational gaps.

To see explicitly how mara can be used to track relatives, it is best to go through an example. Before we do this, however, I *strongly* suggest you grab a scrap of paper (or find a patch of sand) and draw Figure 1 that appears in Chapter 3: it is very hard to think through kinship without it.

To start, let us suppose that I am an individual of type B soon to be married. Following this figure, being of type B, I must have a mother of type A, with whom I share mara, and a father of type D. Further, let's label the mara of my mother and myself as x (which could stand for emu, black snake, grey kangaroo, and so on), and the mara of my father by y (which cannot be the same blood as mine). If I am to marry, from this same figure, the first preference is for an individual of type C. Similarly, this type C individual has a mother of type D and father of type A. As before, let's label their mara and that of their mother as w and that of their father with z. This is all summarised in Figure 2.

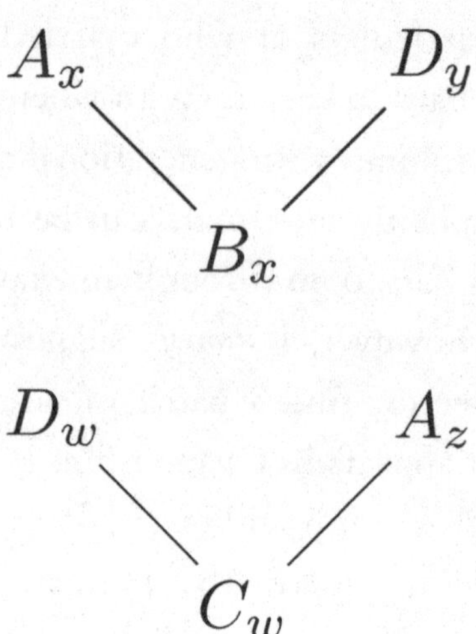

Figure 2: Totem Analysis: the top figure represents the beginning of a family tree for an individual of type B: they share their mara *x* with their mother. Their father, type D, has mara type *y*. The bottom figure represents a potential partner of a type B individual, here labelled C. They share their mara *w* with their mother who is type D. Their father is type A with mara *z*. Note that branches to the left indicate mothers, whereas branches to the right represent fathers. For this reason, and this is *very* important, branches to the left also group blood types: A and B form one pair, and D and C the other.

With this set up, let us see what mara can tell us. First, note that the father of my potential partner (right branch of bottom diagram) shares the same first part of dhii – yibaay, gambuu, marrii or gabii but here labelled A – with my mother (left branch of top diagram). This means that they are the same blood type. If, however, in addition to being the same blood type we also have that $x = z$, that is, that they have the same mara, we know instantly that there is a somewhat close relationship between us. In particular, either my mother and their father are siblings, or cousins through women, or second cousins through women etc. In short, despite potentially very many moves, we know that they share a matriline.

My potential partner's father's father, on the other hand, must be type C.[15] This means he cannot be related to my mother's mother – as they are different blood types – but he might be related to my mother's father who will also be type C. Recall, however, that traditionally men do not migrate and women only do so after marriage. This means that my potential partner still lives with their father's father. If the latter is closely related to my mother's father – who also did

15 Note that this generation has been left out of Figure 2 on purpose. It can be *generated* from an understanding of Figure 1. You will learn a great deal more by doing this yourself.

not migrate – they will necessarily live together. In this case, any relationship between us will be well-known. There is the possibility that their father's father and my mother's father are related through women instead – a generation earlier, through sisters who migrated to different places – but in this case they will have the same mara.

If we now focus on their mother, we know that she has a different blood type to my mother (see Figure 2), and so as above can only potentially be related at the great-grandparent stage, which mara will identify. My potential partner's mother's father, however, has the same first part of dhii – yibaay, for instance – as me. The closest possible relationship that may be here is that their mother's father may be my mother's mother's brother. In this case, he and I will share mara, which one question will reveal.

We have then – using mara – considered every possible way my mother may be related to my potential partner. This is despite any considerable physical distances between our families, or multi-generational gaps. The real cleverness of our totemic system is that the above questions can also be asked from the perspective of the other parents – there is a great deal of what we mathematicians call symmetry.

For instance, if they instead ask what my father's mara is, we will know, very quickly, if he and their mother are related.

For the purpose of traditional matchmaking, as with my grandparents, these questions were asked by, and at the generational level of, their parents. In turn, this meant the mara of each of my grandparents' grandparents could be compared. However, the value of our totemic system, even just at the level of parents, extends well beyond marriage. For instance, knowing only the mara of a stranger's mother and father will go quite some way to uncover any potential links between us. This is why, traditionally, the first thing done upon meeting someone was sharing your dhii: yibaay dhinawan ngaya. Another way to put it is that, while the first part of dhii tells us *how* we should interact with another person – as sibling, or distant mother, or uncle etc., especially when in Western thought they would not be considered as such – mara instead uncovers an additional *why*. This is not, in any way, to downplay the importance of relationships that do not hinge on mara. Instead, it lends to those that do a formal degree of closeness.

At this point, some readers might begin to think of the system of mara as Western last names but through women. This is true, *approximately.* The system of mara also has the much larger and related function of constraining personalised totems (more on this in a bit) – so that all things are cared for – as well as providing a general system of categorisation. This is not unlike the Western field of taxonomy, where all livings things are sectioned into groups. Radcliffe-Brown mentions this briefly in the case of the Wailwan in [15] at the bottom of page 232. Unlike the Wailwan, however, who have two categories, us Gamilaraay instead have three: giinbaligal, dhurrungal and dhigayaa. Respectively, these mean 'scaly tribe', 'furry tribe' and 'feathered tribe'. It is worth cautioning, however, that we did not categorise in the way an English speaker might intuitively: emu sat with the scaly tribe, not the feathered, for instance.

Now, before I touch on the *form* of our totemic system writ large, which was another source of confusion, I will first pause here to address second preference marriage. In the analysis above, the potential partner was considered first preference because they had a different first part of dhii from myself, my mother and my father. Note, from what we have just learnt, that if they shared mara with my father our relationship would be rather close (if $w = y$). This is why second preference marriages exist: on the off chance the above occurs, you may instead marry someone who has a different mara to your father, but same first part of dhii. Note that this means, necessarily, that they will *still* have a different blood type to you and your mother.

Now, because our fathers are unique, and this is an extremely important point, what would be considered a first preference marriage is in fact *dynamic*. More precisely, a first preference marriage for me is not what it might be for other families or indeed generations or even my partner (more on this in the next chapter). This is, I suspect, what caused confusion amongst the purportedly great anthropologists of the 1800s. It is also one of the reasons I opted to be very general, and choose labels A, B, C and D. Despite the added uncertainty that second preference marriages introduce, which in fact compound, the mara system can still successfully track close relations. It is worth going through such an example yourself.

The remainder of this chapter, instead, will be spent attempting to clear some of the fog around form. By form, here I really mean the structure of our totemic system and the levels contained within it. For instance, it may be said that Kaputhin, the eagle, and Dilby, the crow, are totems of all dark bloods and light bloods, respectively. They are, in some sense, emblems. Beneath this level, however, we have the system of mara, the function of which is discussed above. Beneath mara, in turn, there are much more personalised totems that are drawn from fixed lists unique to each mara.[16] To make things more complex, from such lists you may have more than one totem. This means that the average Gamilaraay mari has *at least* three totems, and quite possibly more.

Now, the function of personalised totems is complex, multiple and unsurprisingly highly individual – so much so that I could not do them justice here. Suffice to say these totems grant us responsibility for and relationship with the world around us. This is not true of what we might be tempted to label as totems that are instead protective in nature. Magpies, for instance, are said to guard all Gamilaraay women. This distinction I believe

16 The only published such list, to my knowledge, can be found in [12]. This list is far from exhaustive and it is best, as usual, to speak instead to older relatives. Mathews [9, 10], and more recently Testart [19], have lists of mara by blood but I do not believe these works to be especially useful to Gamilaraay people.

to be very important: protective animals and other totem-like representatives, though spiritually relevant, do not determine the very social and economic organisation of our society. It is worth, in my opinion, being extremely clear about these differences: if we do not, we risk the considered and practical nature of our totemic system once again becoming clouded. In a way, I am trying to highlight how English has betrayed us and continues to do so: terms that would not be confused in our language are instead flattened into the one word. Perhaps, better still, I am suggesting that maybe we ought to discard the word totem entirely, which has meanings beyond even what I've talked about here.[17] If we must use English, instead we might ask: *what is your blood, what is your hand, who do you care for?*

To finish this chapter, it is worth focusing on who we care for. In particular, it feels necessary to return to the earlier point that personalised totems are *constrained* by mara. This fact, I believe, is not appreciated nearly enough. A specific connection – fabricated or otherwise – to an animal, plant, river

17 I have similar feelings towards the word 'clan', used throughout this entire book only once. It is employed in *at least* three different ways in community and in the literature, unsurprisingly leading to a great deal of confusion. I am not at all convinced it serves us well.

or wind does in no way mean that it is your personal totem. If we all simply took the totems we desired, the function of personalised totems vanishes: they create careful networks of responsibility, so that all things on country have someone looking out for them. This is to say that personalised totems, in many senses, are not about us. The Western hyper-individualisation that has crept its way into our totemic system, for this reason, quickly leads to the absurd: desert mob with a dolphin totem serves neither the dolphin nor the locals. Such nonsense should be *gently*, but very seriously, laughed into nonexistence.

A note on adoption and rejuvenating mara

Recall how, in Chapter 4, I mentioned that people who have Gamilaraay fathers, but not mothers, will need to be traditionally adopted if their mother does not also come from a four-group kinship system. Some reasons, I hope, are now beginning to become clear. The first is that the blood type of hands, due to inheritance rules, does not change: dhinawan has been dark blood since Gamilu wiibidi. If we begin to not just remember, but also take our father's hand, this need not remain true. The second reason is that if some of us begin to inherit our father's mara or indeed blood as our own, marriages in the past that would have been forbidden may become permissible, and marriages that would have been encouraged are no longer allowed.

As a concrete example, consider a man and a woman who have not been closely related for centuries but who, due to the particulars of history, have the same mara. Let's call this mara brolga. Now, this woman's daughter's daughter, if we follow matrilines, will have the same blood as her grandmother and the same hand, brolga. If the man passes his mara on to his daughter, who then passes it on to her son (i.e. if they do not strictly follow matrilines), this grandson will be considered brolga. In this case, he cannot marry the woman's granddaughter. Note, however, that if a matriline had been followed instead, the man's daughter would have a different mara and blood type to her father. In turn, her son would inherit that mara and blood type. In other words, traditionally, he would have been a perfect candidate to marry said granddaughter.

The instance where a forbidden marriage becomes possible is simpler still, and therefore more worrying: a dark blood woman has dark blood daughters, whereas light blood men – if we follow matrilines – instead make dark blood sons. For this reason, the children of dark blood women and light blood men are traditionally forbidden from marrying. If, however, a light blood man's sons take his blood, it would seem that the above marriage is permitted. In even the very recent past, such a union would have been frowned upon.

For some, such traditional adoption to fix these problems will seem very odd. For others still, it will be a difficult exercise in the letting go of pride. I want to confirm that these practices are nothing new to our people and our lands. I also want to confirm that, so to speak, it is not just men marrying out of kinship that creates situations we must think deeply about: though my siblings and I were not born handless, we were strictly speaking born without land. This is because, while three of our grandparents were Indigenous, our father's father – the line through which Gamilaraay land is inherited – was not.

We have been, instead, taken in by our maternal grandmother's country. The point, I suppose, that I'm trying to make is that despite the fact our mother's aunties, grandmother and many ancestors before them are buried there – and we remember exactly where – that land was not our inherited right. The notion of being Gamilaraay was not enough, either: to be taken into that land required grace, generosity and permission. It also required, before all of that, an acceptance of what was not ours.

DHULANBAA V

1.
This will be difficult to express in the language of my coloniser. I will fail.

2.
I will try.

3.
If you pound the bark of the Black wattle, a powder forms that is toxic to fish but not to us.

4.
This is why, all along waabi's country, there are billabongs that make no sense: detours of the river moulded by hands.

5.
In these no-sense billabongs, this powder stuns the fish. Fragile and dopey, they float to the top.

6.
We take the fish we need,
we leave the fish we need.

7.
A question: does the pounded bark make the yamstick
made from wattle work less well?

8.
Kinship is the same, it can have more than one use.

9.
Kinship is the same, it can be poison to me but not
to you.

10.
By this I mean, Western ways of relating have not
served us Gamilaraay well:

we are sick.

11.
Exhale: around this bend, the main river runs on.

5

Gamilaraay kinship terminology and its relationship to dhii

IN GAMILARAAY KINSHIP, THERE IS a push and a pull. The system of mara, discussed in Chapter 4, is the push that ensures we do not get too close. The system of dhii on the other hand – the focus of this chapter – gathers us together: from this pull flows, in fact, a whole host of complex kinship terms. These terms classify, in a way that is counter-intuitive to most, all Gamilaraay people as some type of family member.[18] For instance, an unrelated Aunty I've just met might be categorised as my father's mother, or the child of a stranger might be considered my son. This reaching out, and folding in, ensures that no man is an island; a blueprint for how we ought to relate to others, as well

18 And, remarkably, other mob further still.

as our responsibilities to them, is hardwired into our kinship system. This chapter will be a very shallow attempt at explaining this. Mostly, however, it will be a call to other Gamilaraay mari to think deeply and *often* about our kinship terminology.

While this way of relating is obviously an important function of dhii, I must be very clear here before progressing: this is not its sole purpose, either historically or perhaps in the future. It is not hard, for instance, to imagine a modern system of election whereby yibaay/yibadhaa votes in yibaay/yibadhaa, gabii/gabudhaa votes in gabii/gabudhaa and so on, to create a council of four individuals who decide on important Gamilaraay matters. Whether we innovate to create such a governance system remains to be seen. The point I am making is that all of our old systems brim with potential: we should not restrict ourselves due to some false notion of a static, unchanging society, pre-colonisation. All of that said, the practical day-to-day way of relating developed by our ancestors is an inheritance not to be taken lightly; this chapter, then, will feature very little innovation. Instead, it will be a small step towards rejuvenating the daily usage of kinship terms.

Now, despite my many misgivings, the best place

to start discussing Gamilaraay kinship terminology is in fact the Yuwaalaraay-Gamilaraay dictionary.[19] The reason it's the best place to start is that very many kinship terms here are collected, which as we will see both helps and hinders the Gamilaraay learner. For instance, if we look up the words for 'grandmother' we find the following: miimii (YR), mamaay (GR), baagii (YR, YY), ngaagii (YR, YY), waabi (GR), badhii (GR), gaadhii (YR), and garrimaay (GR, YR, YY). Nganga (GR) is also listed in Austin [1]. In this list above, GR indicates that the word was recorded in Gamilaraay, YR in Yuwaalaraay and YY in Yuwaalayaay, a closely related language.

All of these bar one (miimii) are listed *at least* as either 'father's mother' or 'mother's mother', with no surprises there. Miimii, in fact, is noted as being quite possibly a loan word from Wangaaybuwan to the south. Things become, however, quite muddled when we compare the remaining words: all of them recorded with Gamilaraay origin are listed not only as a particular grandmother or her sisters, but also her brothers. This is not true of the Yuwaalaraay and Yuwaalayaay entries.

19 The online dictionary is, to be clear, a brilliant resource and for that I am grateful. That does not mean, however, that Gamilaraay people cannot want it to be better.

For instance, 'waabi' is listed both as 'mother's mother' and 'mother's mother's brother'. Indeed, the dictionary entry reads:

> This is a rare word, the common word is baagii. It is an unusual kinship term because it refers to a female and male. It may be wrongly recorded.

For 'badhii' we find something similar:

> Sources also give this as 'mother's mother's brother', but it is unlikely that one word means both 'mother's mother' and 'mother's mother's brother'. This is a rare word.

The word for 'father's mother', garrimaay, is also listed as 'father's mother's brother' in Austin [1]. Here, 'nganga' is given as an alternative. In the dictionary proper, garrimaay is also listed as 'mother-in-law' and 'son-in-law', the reason for which will become clear in a bit.

Now, presented with this data we must, going forward, accept one of two things. The first is the possibility that the linguists are correct, and that such kinship terms could not refer simultaneously to a grandmother, her sisters *and* her brothers. The second is that the linguists are wrong. This, in turn, would have a very surprising consequence: Yuwaalaraay and Gamilaraay people, who share many customs and have mutually intelligible languages, would necessarily then have fundamentally different kinship systems. This surprise would be elevated further by the fact that we share even the same words for dhii: yibadhaa, buudhaa, gabudhaa, maadhaa and so on. In this sense, the linguists – who I do in fact believe to be wrong – deserve some grace.

The key to understanding our kinship differences lies with the following observation: the Yuwaalaraay – if we are to accept early accounts – did not *historically* have second preference marriage [12, 15]. In other words, yibaay/yibadhaa always married gabudhaa/gabii, while gambuu/buudhaa always married maadhaa/marrii. Though this detail seems small, it has huge implications for who can be classified with whom under the system of dhii. For instance, with only first preference marriage, my father's father will *always* have the same dhii as my mother's mother's brother. This is reflected in the Yuwaalaraay term 'dhilaagaa', which has precisely both meanings. When second preference marriage is permitted, as with the Gamilaraay, dhii is not preserved in the same way. In turn, the daily terms used to refer to people shift and flow.

Following the consequences of second preference marriage on dhii is a little involved, so I will go through it here. First, however, some shorthand: in what follows 'mm' will refer to 'mother's mother', 'ff' to 'father's father', 'mf' to 'mother's father' and 'fm' to 'father's mother'. In addition to this, as before, let's assume that I am an individual of type B. Further, let's assume my mother has an immediate family history that coincides with the left diagram in Figures 3, 4, 5 and 6. Namely, not only does she have a mother of type B but her father is type C. A first preference marriage for her is then to a person of type D.

With that set up, we can now turn our attention to Figure 3, which represents the historical account of Yuwaalaraay marriage. The left diagram represents a B type individual's mother's immediate family history, while the right diagram represents that of their father. As there is no second preference marriage, all individuals have what would be considered first preference marriage under Gamilaraay kinship. Tracing through matrilines, we find the following types on the matrilineal side (left diagram): 'mm': B;

‘mf’: C. Again, tracing through matrilines, but this time on the father’s side (right diagram) we have: ‘ff’: B; ‘fm’: C. Now, what is true of an individual is also true of their siblings if they share a mother. In turn, this means that the brother of ‘mm’ is also type B, the sister of ‘mf’ is also type C and so on. In this way, relational words such as ‘dhilaagaa’, as mentioned above, naturally arise.

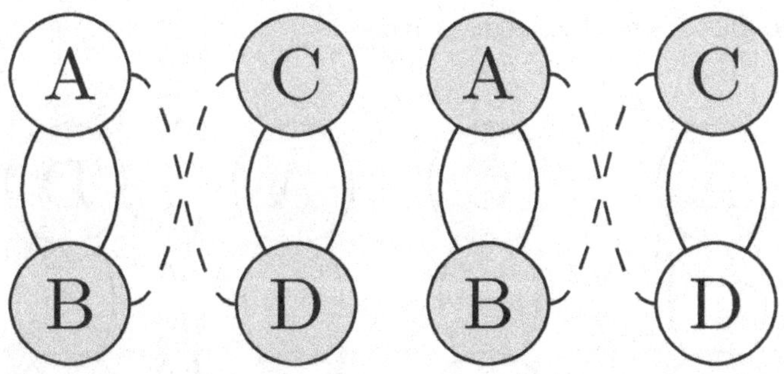

Figure 3: Left and right diagrams represent a B type individual’s mother’s and father’s immediate family history, respectively, with white circles indicating those parents. This instance represents first preference marriage for both parents.

Figure 4 on the other hand includes second preference marriage, and so a divergence from the *historical* Yuwaalaraay system appears. Importantly, however, note that it represents an instance where the marriage (in this case, between a type A mother and type C father) is first preference for one parent but not the other. In particular, it is first preference for the father but second preference for the mother. In this case, tracing on the mother's side (left diagram) we find: 'mm': B; 'mf': C. Similarly, on the father's side (right diagram) we now have: 'ff': B; 'fm': D. Here already we see that 'mf' and 'fm' (and so too their siblings) do not have the same dhii.

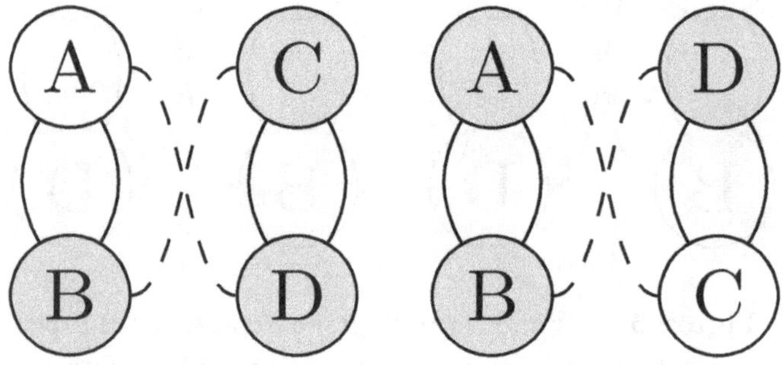

Figure 4: Left and right diagrams represent a B type individual's mother's and father's immediate family history, respectively, with white circles indicating those parents. This instance represents second preference marriage for the mother but first preference for the father.

Figures 5 and 6 represent the remaining two possibilities. Namely, the instance where marriage is first preference for the mother and second preference for the father (Figure 5) and second preference for both parents (Figure 6). In the first case, we find: 'mm': B; 'mf': C; 'ff': A; 'fm': C. In the second case, instead we find: 'mm': B; 'mf': C; 'ff': A; 'fm': D.

Observe that in all four cases it is only 'mm' or 'mother's mother' that reliably stays the same: this is yet another reason why – in the regrowth of our kinship – the following of matrilines is so important.

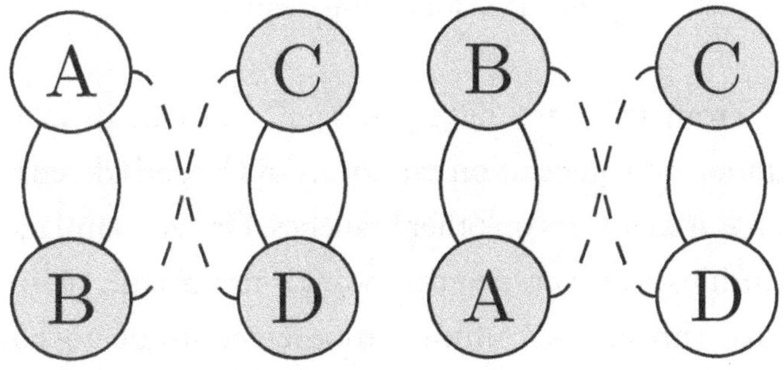

Figure 5: Left and right diagrams represent a B type individual's mother's and father's immediate family history, respectively, with white circles indicating those parents. This instance represents first preference marriage for the mother but second preference for the father.

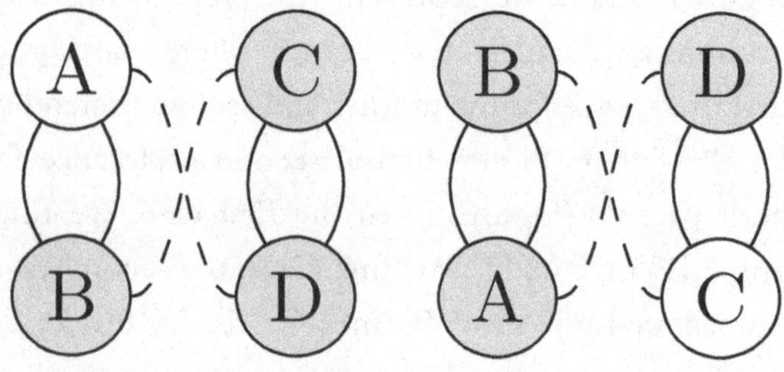

Figure 6: Left and right diagrams represent a B type individual's mother's and father's immediate family history, respectively, with white circles indicating those parents. This instance represents second preference marriage for both parents.

From what we have just seen, individuals at the grandparent generation cannot always be considered the same as members on other branches. Under Gamilaraay kinship, however, and this point is important, siblings born through a common mother continue to have the same dhii. An obvious consequence of this is that your grandparents will share dhii with both their sisters *and* their brothers. This means, in turn, that terms such as 'waabi', which were recorded as both your 'mother's mother' and her 'brother' are not so surprising. Indeed, Radcliffe-Brown referred to social organisation around these particular categories as Arrernte[20] type kinship [14], which is present on vast swathes of the continent. The other canonical kinship type that Radcliffe-Brown identified (though there are, in truth, more than two) instead categorises individuals as historical Yuwaalaraay do [15].

Though I may be the first to link the presence of second preference marriage with a change of kinship system, I am not, by a long way, the first to note that Gamilaraay kinship is Arrernte type: Radcliffe-Brown [15] and Austin, by way of Tindale's work in the 1930s [1], stated this in no uncertain terms.[21] This renders the discouragement of the use of certain Gamilaraay terminology noted above (i.e. 'This is a rare word'; 'May be wrongly recorded'; 'Unlikely') as either academically negligent or as evidence of the active colonial erasure I attempted to highlight in Chapter 2.

Now, the purpose of this chapter is not to bemoan any harm committed against us – intentional or otherwise. The aim instead is very practical. The rest

20 Note he used the alternative spelling 'Aranda'. Arrernte is the more common spelling used now by Arrernte people themselves.

21 These things, I'm sure, were well understood prior to invasion: first, second etc. here are laughable timescales in the face of the very long history of Gamilaraay knowledge production.

of this chapter then, with help from Arrernte kinship, will be spent building a guide of Gamilaraay kinship terminology with a focus on understanding its *dynamic* relationship to dhii. To this end, I will refer heavily to the exceedingly generous work of Arrernte elder Veronica Perrule Dobson, *Anpernirrentye Kin and Skin: Talking about family in Arrernte* [3].

In this work, Dobson confirms that the words 'arrenge' (ff), 'aperle' (fm), 'atyemeye' (mf) and 'ipmenhe' (mm) refer not just to individuals but also their siblings. For instance, just as with 'waabi', 'ipmenhe' can be a reference to your mother's mother, her sisters or her brothers. This provides strong support for future Gamilaraay usage in a similar way. In addition to this, and with great clarity, Dobson notes the reciprocal nature of these kinship terms. For instance, 'atyemeye' means not just 'mother's father' but also a '*man*'s daughter's child'. In other words, the grandparent and the grandchild refer to each other with the same word.[22]

This is reflected too in the Gamilaraay records, with the dictionary entry for 'dhaadhaa' being listed

22 It will not be lost on any Gamilaraay readers that this persists in our variety of Aboriginal English: our toddlers very often get called granny, poppy, aunty etc. by the very people who have those terms.

as both 'mother's father' and 'daughter's son'. This entry, however, if we pay attention to Arrernte kinship, seems to lack nuance. If we follow the Arrernte pattern, it would instead have the larger meaning of 'mother's father' and a 'man's daughter's child', not just her son. Note that it wouldn't mean a 'women's daughter's child', which is 'waabi'. The reason such detail was likely not recorded is almost certainly due to the cultural biases of the early anthropologists themselves: more often than not, they asked questions of men and not women, so that the people who told them 'dhaadhaa' meant 'daughter's son' were very likely, and to themselves self-evidently, men. If anthropologists had consistently asked their wives the same question, I suspect this intricacy might have been recorded.

To pull things together, waabi (mm), garrimaay (fm), dhilaa (ff) (see Austin [1]) and dhaadhaa (mf) can mean not just these respective individuals, but also their siblings. More than this, these terms can also refer to grandchildren of a specific type: if a child – irrespective of gender – refers to a grandparent with a specific title, the grandparent and siblings call them that same title. For instance, I refer to my mother's mother as waabi, as well as her brothers. They, in turn, if we follow the Arrernte pattern, refer to me and my siblings by the same name. On the other hand, my mother's mother and her brother refer to his grandchildren with a different term, which will be reciprocated. The key here is that dhii, which is shared by siblings, *determines* relationships.

Moving down to the level of parents, much is known and still regularly used: 'bubaa', for instance, meaning 'father' or 'father's brother', can be heard in many Gamilaraay families. 'Gunii', 'gunidjarr' and 'ngambaa' are also in regular use to mean 'mother' or 'mother's sister'. These conform, unsurprisingly, with the Arrernte pattern. Similarly, 'father's sister' is well known to be 'walgan' or 'baaman', while 'garruu' is still used to mean 'mother's brother'. These terms, however, have additional meanings that can be counter-intuitive to English speakers. For instance, 'father's sister' is also listed in the dictionary as 'mother's brother's wife' *because* these two (potentially different) individuals have the same dhii: they are *classificatory* sisters. By symmetry, 'garruu' then can also mean 'father's sister's husband'. Again, this is because of dhii: if your parents are right for each other, then so too are their siblings.

At this generation there is still some confusion,

however, surrounding terms relating to in-laws. For instance, 'mother-in-law' is sometimes recorded as 'walgan', other times it is listed as 'garrimaay', which you'll recall from above means 'father's mother'. On first glance, this suggests something is amiss: perhaps the person recording got things wrong or perhaps the person recorded didn't know. Often this last explanation is relied on, calling into question how well we Gamilaraay know ourselves. The answer – as with all puzzles relating to our kinship – lies with dhii. In particular, if you have a first preference marriage your mother-in-law will be a classificatory walgan; whereas if you have a second preference marriage she will be a classificatory garrimaay. This can be confirmed for fathers, as shown earlier in Figure 4, and mothers in Figure 5.

A similar thing happens with fathers-in-law; however, in this case the change occurs according to whether or not your partner had a first or second preference marriage, not yourself. Again, this can be confirmed in the same figures. More precisely, if your partner had a first preference marriage, your father-in-law will be a classificatory garruu, which is recorded both in the Yuwaalaraay-Gamilaraay dictionary and Austin [1]. On the other hand, if your partner had a second preference marriage then your father-in-law will share dhii with you. It might be tempting to then label him as 'dhilaa', though an alternative listed in the dictionary is 'bambuy'. This highlights an important quality in Gamilaraay kinship: every single individual has more than one way that they relate to you, and the way you refer to them will depend on context.

Now, just as garrimaay is reciprocal between grandparents and grandchildren, a similar use has been recorded for in-law relationships. In particular, 'garrimaay' is listed not only as a certain type of mother-in-law but also a 'woman's son-in-law'. Given all of the dhii considerations above, however, this term can also mean a 'woman's daughter-in-law' *when* dhii creates the correct classificatory relationship. The reciprocal nature of this term then almost certainly comes from the fact that it is a grandparent relation and not because it is an in-law one. Similarly, 'bambuy' is recorded also as a 'man's son-in-law', but for the same reasons might also mean 'daughter-in-law' when dhii permits.[23]

When father-in-law and mother-in-law instead

23 This will become much clearer after reading the next chapter, but in essence second preference marriage has the effect of skipping a generation, which then creates reciprocal terms.

are classificatory garruu and walgan, it makes less sense that the terms are reciprocal, though Austin does have 'garruu' listed in that way. The reason is that the reciprocal terminology above arises because of what is called generational moieties; in Arrernte, this is captured in the term 'nyurrpe' [3], in Yuwaalaraay/Gamilaraay by 'manday' [7]. The crux of this idea is that you are, in some sense, the same as your grandparents and grandchildren because it is at this stage that kinship, in a way, resets. Following matrilines, yibaay and yibadhaa, for instance, becomes once again yibaay and yibadhaa only when you reach the grandchild stage. More concretely, my siblings and I are yibaay and yibadhaa, the same as our waabi, my mother's mother and her brothers. By contrast, the terms for son-in-law and daughter-in-law in the garruu and walgan cases will be those classified as a certain type of niece and nephew.[24]

All of the terms explained above, I hope it is becoming clear, are not just different words but

24 Note that in Arrernte only one of the set of four in-law terms arises in a classificatory way: 'ahenterre' or 'ampatye-ampatye' (meaning classificatory child) is listed as a 'woman's father-in-law', whereas 'meye' (which also means 'mother') is given for a 'man's daughter-in-law' [3]. While the other in-laws are also classificatory relations – everyone is related – there are additional terms to mark the in-law quality. Gamilaraay has either lost this, or we didn't have them to begin with.

different *concepts*. The closest Gamilaraay equivalents of niece and nephew drill this point home especially well: it is not simply a matter of translating but of *thinking differently*. To see this, we must return again to dhii: note that, owing to matrilineal inheritance, my sister's children will have the same blood as me, whereas my brother's and my own will not. In particular, my sister's children are buudhaa/gambuu (like my mother and her siblings); whereas mine will be, depending on the type of marriage I have, either marrii/maadhaa or gabii/gabudhaa. In turn, this means the way that my sister and I, who are yibaay and yibadhaa, *both* refer to her children will be different to how we both refer to mine. Niece, nephew, son and daughter are, in fact, very poor approximations to what we have in Gamilaraay.[25]

Austin lists 'sister's daughter' as 'gunugayngaa' and 'sister's son' as 'gunubingaa', and noted that these were

25 The switched-on reader will note that while all three of my sisters' children will have the same dhii (due to matrilineal inheritance), it is in fact theoretically possible that my children and those of my brother might have dhii different to each other. This will arise if my brother has a first preference marriage and I a second or *vice versa*. In such a case, my brother's children will have dhii that coincides with that of my daughter's children: they, counter-intuitively to English speakers, will have the label of a grandchild *from my perspective*. While in the past this was very unlikely to occur due to groups of siblings (understood through dhii) being promised to each other, this highlights the nuance needed when thinking through Gamilaraay relations. It is a mistake to simply map to Western concepts.

recorded from the perspective of men [1]. 'Son' and 'daughter', likely also recorded from the perspective of men, are listed in the dictionary proper as 'wurrumay' and 'ngamurr', respectively. If we follow the logic of the above, and the pattern of Arrernte, both my sister and I will call her children 'gunugayngaa' and 'gunubingaa'. On the other hand, we will both refer to my children as 'wurrumay' and 'ngamurr'.[26] While this may seem confusing at first, it is in fact partially still practised in our variety of Aboriginal English: because of dhii, my mother's sisters consider my siblings and myself as their children, while we consider them to be our mothers.

All of that said, it is important to impress that we Gamilaraay are not static: we have and continue to change. 'Miyay', 'birraay' and 'gaayli', once meaning 'girl', 'boy' and 'child', respectively, are now commonly used also to mean 'daughter', 'son' and '*my* child'. This son and daughter, however, are son and daughter as understood from the Western perspective, not the more traditional Gamilaraay one.

The final generation to consider is that of our own;

26 In turn, following dhii, when father-in-law is 'garruu', daughter-in-law and son-in-law will be the words I use for my sister's children. Conversely, when mother-in-law is 'walgan', daughter-in-law and son-in-law will be classificatory 'ngamurr' and 'wurrumay'.

the people we call siblings, cousins and lovers. It is, having understood all that's been said so far, fairly straightforward. 'Guliirr', for instance, is the gender neutral term for either 'husband' or 'wife'. By contrast, words for siblings are gendered but their seniority is much more important – 'baawaa' is an option for 'older sister', while 'bariyan' is used for 'younger sister'. 'Dhaya' is the masculine version of 'baawaa', while 'galumaay' is found in place of 'bariyan'.[27]

For those paying attention, the reason I very particularly said seniority, and not age, is because your mother's sister's children, who share dhii with you, are also considered siblings. If we follow the Arrernte pattern, the seniority of these slightly-more-distant siblings depends on the relative ages of your guniigalgaa or 'mothers', not of you and them.

This concept of near or far siblings is in fact very important for all aspects of Gamilaraay kinship: all people of your generation with your dhii are siblings, near or far. All women of your mother's dhii and generation are your mothers, near or far. All people of your father's mother's generation and dhii are your garrimaay, near or far, and so on. The point is that dhii,

27 While 'dhagaan' is commonly used today to mean any 'brother', it previously meant another person of any gender with the same dhii. See dictionary entry.

which determines relationships, does not see genetic closeness: that is the role of the system of mara.

This is not at all to say, however, that dhii does not *affect* genetics: it determines average genetic relatedness of couples very cleverly as a function of population size [4]. This, in turn, reduces average incidence of recessive diseases. This is no doubt an artefact of the fact that Gamilaraay kinship was *constructed.* K. Langloh Parker, who had a cattle station at Narran Lakes for some time, went so far as to say:

> Blacks were early scientists in some of their ideas, being before Darwin with the evolution theory, only theirs was a kind of evolution aided by B___. I dare say, though, the missing link is somewhere in the legends. I rather think the Central Australians have the key to it. One old man here was quite an Ibsen with his ghastly version of heredity [12].[28]

Genetics aside, the last group to consider are those cousins with whom we do not share blood or dhii: they are those born through your mother's brother and your father's sister. The term for this type of cousin,

28 Note that the word above has been censored because it is taboo to speak for the vast majority of people.

on the face of it, we have lost in community. Worse still, it appears that such a loss occurred by the time linguists reached our lands. The solution, however, as always lies in dhii: those cousins are the one and the same, mara permitting, who you are allowed to marry. Otherwise put, your potential partner is a classificatory cousin. Or, thought differently again, your cousin of the above type is a classificatory potential partner, even though mara would exclude such a marriage. The word, then, for the potential wife of a man, 'nganuwaay', might also be used for female cross-cousins.[29] Similarly, the term for a sister's husband, 'biraman', is an option for male cross-cousin.

Note, however, that these terms (like most, unfortunately) were apparently recorded from the perspective of men. From the perspective of women, 'gambaay' has been recorded as sister-in-law, and so too is a candidate for a woman's female cross-cousin. A woman's potential husband, on the other hand, has not been recorded. There are, however, many classificatory options (including those that are gender neutral) that might fill this gap. We must however, as

29 A word of caution: while a female cross-cousin is always a classificatory nganuwaay it is not true, due to second preference marriage, that a nganuwaay will always be a female cross-cousin. This is why going forward Gamilaraay people must draw our understanding from *structure*, first and foremost.

a collective, be careful here to not fall into the same old anthropological traps. In Arrernte, for instance, while there are separate words for brother-in-law and male cross-cousin, even though they can be classificatorily the same person, none of those terms change as a function of the gender of the *speaker.* This suggests that the terms listed above might have had broader meanings than were originally recorded. For example, it might have been that once both biraman and gambaay were used by either gender, as is done in Arrernte. Community as a collective will, *amongst ourselves*, need to settle on which terms feel most natural.[30]

At this point, it is worth reflecting on what we have learnt. By this, I do not mean the nitty-gritty of particulars, or even a general summary of them. I mean instead the seeing of the image as a whole: the system of dhii, in very concrete terms, gathers us together. Of the many thousands of Gamilaraay people currently alive, I am linked to every single one of them in a *usefully, precisely and closely* defined way. I have responsibilities to them, and them to me,

30 Anticipating the disappointment of this last sentence, I note that *anybody* can be referred to or called by their meat name: for instance, Yibaay Dhinawan is both an informative term and correct name for me. The only thing it lacks is generational information, which speakers can infer contextually.

even if we have not yet met. Through dhii we are, without exaggeration, family. The intention here is not to invoke vague feelings of warmth: families are equally sites of support as they are of uncomfortable accountability. The intention instead is to highlight the following: through dhii, in numerous ways, we are rendered stronger.

A caution on labels

At the risk of extending this chapter well beyond its intended length, I want to take a moment here to provide a warning: there are many words in community that are not found in the dictionary or Western scholarship, more generally. For this reason, many of the above *labels* (baawaa etc.) may not be used by certain families. Indeed, they may have kept entire sets of different words to the ones provided here. To my mind, that is a very good thing: colonisation sought to destroy our culture but we have all kept different parts of the puzzle. In no way, however, does having different labels contradict the *structure* of our kinship that I am trying to bring to light. Indeed, it enriches it.

DHULANBAA VI

1.
This will be difficult to express in the language of my coloniser. I will fail.

2.
I will try.

3.
Dhulanbaa is the time of the Black wattle.
Dhulanbaa is the place of the Black wattle.

4.
There is no time without the place, and no place without the time.

5.
We do not mark each beat and swallow it whole.
We are rhythm people.

6.
When you die, said Garruu, you may be reborn not as person or animal. Not even as tree.

You may, instead, be reborn as place.

7.
This makes mining murder.

8.
This also makes us strong.

9.
When I say we are rhythm people I do not mean that we just track them. And certainly not force them.

What I mean is that we *participate* in a rhythm that is larger than ourselves.

10.
Those who do not may perturb it, *yawu*. But they cannot break it.

Gamil.

11.
Take solace, my people. Rest easy.

12.
Wake early, my people. There is much to do.

13.
This will be difficult to express in the language of my coloniser. I will fail.

14.
I have tried.

6

Kinship, locatedness and death

THIS BOOK IS A BEGINNING. For this reason, there's no real proper way to end it. Many topics, for instance, come to mind – how kinship maps to land, how non-human relatives are included, how non-living phenomena, such as the winds, are accounted for, too. These topics, however, I leave to other Gamilaraay thinkers; this book is a beginning. Instead, here I will attempt to collect core ideas throughout the chapters and, so to speak, tie off the weave.

To start, I must first make a slight apology. In the previous chapter, I was not entirely honest. There I said that there is an equivalence relationship between grandparent and grandchild – arising from dhii – that affects both how we relate and refer to each other. This part is certainly true. However, this dhii equivalence is not the only one: there are, in fact, more equivalences than there are generations.

This is best understood by referring to Figure 7. Following matrilines, dhii also creates an equivalence between great-grandparents and children, between great-grandchildren and parents, and so on and so forth. There is, to be clear, a continuous cycle that connects past generations with those yet to come, and which extends as far as we have the time to think about it. This cycle is, in fact, very practical: in the same way that terms for grandparents and grandchildren are shared, so too are the terms used to refer to these other groupings. If I ever meet my great-grandchildren, for example, I will call them as I do those in my parents' generation.[31,32]

---

31 For how this is done in Arrernte, it is worth, as always, referring to [3]. Though we Gamilaraay are matrilineal, instead of patrilineal, this book is invaluable to anyone interested in kinship.

32 Note, from Chapter 5, that at the parental generation there are four types of people: gunii, garruu, bubaa and walgan. Assuming first preference marriages, it is straightforward to check that each of your great-grandchildren will fall into the dhii categories of these relationships relative to you. For instance, if you are gambuu, your great-grandchildren will be, like your mother, mother's brother, father and father's sister, yibaay/yibadhaa or gabii/gabudhaa.

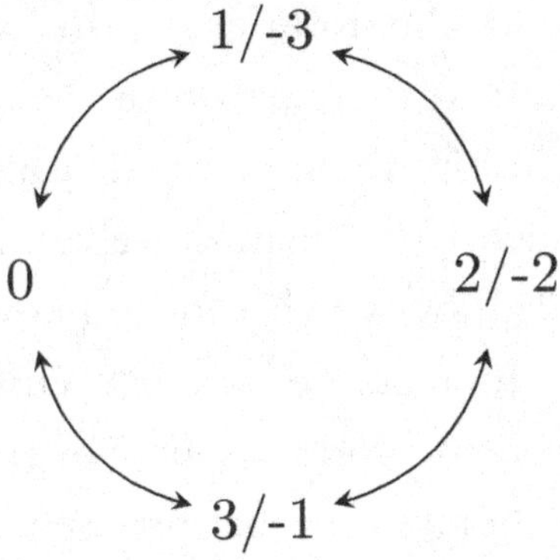

Figure 7: Generation cycles: 0 indicates the current generation with movement clockwise representing previous generations so that 1 signifies parents, 2 grandparents, 3 great-grandparents and so on. Movement counterclockwise, on the other hand, represents proceeding generations: -1 is children, -2 grandchildren, -3 great-grandchildren and so on. Horizontally and vertically opposite groups have reciprocal terminology: just I as refer to my grandparents with the same terms, so too do my children and my parents. Note also the way that generations above and below grandparents and grandchildren, respectively, wrap around: great-grandchildren are, for instance, the same as my parents, which will be reflected in kinship terminology.

Unsurprisingly, this longer cycle of kinship has deeper consequences, too. The most important, in my view, is the following: this cycle *creates* locatedness, just as the rhythm of the wattle does. Existing in kinship, then, grants us participation in something much larger than ourselves. In turn, the feeling provided, by which I mean knowing, is one of a constant, unbreakable *returning*. This is, in what feels like a turbulent world, a source of great comfort.

Before I elaborate on what precisely is meant by returning, first let me clarify something else. When I say that we are granted locatedness, I mean that we become participants in a repeating, predictable cycle that has *both* a time and place element. We are, beyond the timescales of our lives, recurring events. This might seem an odd way to characterise people, but it is very natural once you realise that location in kinship provides both necessary elements at once.

The kinship system itself, on the other hand, lays out the pattern to be followed. More concretely, a yibaay man with a buudhaa mother of the dhinawan mara (the place, position in kinship) will surely occur again, at the very least, through my sisters' daughters and their sons (the time, which *particular* Gamilaraay mari you coexist with in kinship).

Returning, gleaned from what was just said, might then be interpreted to be a sort of reincarnation. These beliefs certainly exist amongst many Gamilaraay. However, in the present context, it is not quite what I am trying to get at. The constant returning mentioned above preferences the collective over the individual: the location that I occupy right now – Yibaay Dhinawan – has been and will be again; I am, so to speak, keeping the seat and the name warm. This is to say that the location itself, and the relationships it grants, are much more important than any person's unique experience of it.

In all of what has been said so far – reincarnation, generational cycles, large timescales and so on – quietly lurches death. Through the intermediary of land, decomposition and regrowth it allows the cycle to trundle on. Flipped on its head, however, our kinship system instead might be viewed as a way to cheat it, to ensure that Yibaay Dhinawan, with the particular responsibilities and relationships that only Yibaay Dhinawan has, carry on as usual. For this reason, the tragedy of death, to the Gamilaraay mind, might arguably not be the bodily death itself but a relocation of presence. There is, however, solace here too: though I will not meet my garrimaay again, another Yibaay Dhinawan will.

Whether or not the presence that has been

relocated returns in exactly the same form is irrelevant.[33] It matters not to me that the formational experiences I have had – that shape me – might not shape any of the next Yibaay Dhinawan. What is important is that the ones who follow exist with the same relationships – Buudhaa Dhinawan gunii, Gambuu Dhinawan garruu, Yibadhaa Dhinawan waabi, Yibadhaa Dhinawan baawaa and so on. Us dark bloods, however, cannot do this alone: we need other Gamilaraay families to remember who they are, and propagate it forward. In this sense, this book is a plea: tend to the Eagle and the Crow, so that the Eagle and the Crow may tend to us.

In Dhulanbaa V, I said that kinship can be poison to me but not to you. This was far from a throwaway comment: by saying that I meant that Western ways of relating have not served us well. I'd also question if the nuclear family – one mother, one father and two or so kids – serves Westerners either. Under the Eagle and the Crow, by contrast, every single relationship you can think of has more than one person filling the role. The consequences of this are obviously huge: the support you both offer and receive is multiplied,

33 Note that some mob, such as the Arabana for instance, once believed, in fact, that we change genders, dhii and even mara each time we reincarnate [8] p. 216.

many times. Regrettably, the practice of this specific way of relating is less and less common. Few of us have more than four grandparents, more than one mother or father, scores of siblings, cousins, children, grandchildren. The Eagle and the Crow, however, is the map to find our way back, to return to a stronger Gamilaraay. This is all to say that if kinship can be poison, it can be medicine, too.

Bibliography

[1] Austin, P. *A Reference Grammar of Gamilaraay, Northern New South Wales*, La Trobe University, Department of Linguistics, Melbourne, 1993.

[2] Bell, R. 'Bell's Theorem (Reductio ad Infinitum): Contemporary Art—It's a White Thing!', *e-flux Journal*, issue 129, September 2022. Accessed: 2023-02-08.

[3] Dobson, V. and Henderson, J. *Anpernirrentye Kin and Skin: Talking about family in Arrernte*. IAD Press, Alice Springs, 2013.

[4] Field, J.M. 'Gamilaraay kinship revisited: incidence of recessive disease is dynamically traded-off against benefits of cooperative behaviours', *arXiv preprint arXiv:2104.14779*, 2021.

[5] Fison, L. and Howitt, A.W. *Kamilaroi and Kurnai: group-marriage and relationship, and marriage by elopement, drawn chiefly from the usage of the Australian Aborigines: also the Kurnai tribe, their customs in peace and war.* Melbourne, George Robertson, 1880.

[6] Gardiner-Garden, J. 'Defining Aboriginality in Australia', Department of the Parliamentary Library, 2003. Accessed: 2022-10-10.

[7] Giacon, J. and Sim, I. *Yuwaalayaay, the language of the Narran River.* Walgett, NSW. J. Giacon, 1998.

[8] Malinowski, B. *The Family Among the Australian Aborigines: a sociological study*, vol. 2. London, University of London Press, 1913.

[9] Mathews, R.H. 'The totemic divisions of Australian tribes', *Journal and Proceedings of*

the Royal Society of New South Wales, vol. 31, pp. 154–176, 1897.

[10] Mathews, R.H. 'Kamilaroi Divisions', *Science of Man*, vol. 1, pp. 155–158, 1898.

[11] Munro, L. 'Radical Black Futures' interview for Sydney Writers Festival, November 2021.

[12] Parker, K. Langloh. *The Euahlayi Tribe: a study of Aboriginal life in Australia*, vol. 1. Library of Alexandria, 1905.

[13] Parker, K. Langloh. *Australian Legendary Tales: folk-lore of the Noongahburrahs as told to the piccaninnies.* London, D. Nutt; Melbourne, Mullen and Slade, 1897.

[14] Radcliffe-Brown, A.R. 'The Social Organization of Australian Tribes', *Oceania*, vol. 1, no. 1, pp. 34–63, 1930.

[15] Radcliffe-Brown, A.R. 'The Social Organization of Australian Tribes', Part II. *Oceania*, vol. 1, no. 2, pp. 206–246, 1930.

[16] Ridley, W. *Kámilarói and Other Australian Languages.* Thomas Richards, Government Printer, Sydney, 1875.

[17] Spearim, P. 'Garruu (Uncle) Paul Spearim (Winanga-li Gii) of the Gamilaraay Nation Tells the Story of Buwadjarr and the Creation Story Part 2', *The National Indigenous Times*, vol. 13, no. 342, p. 18, 2014.

[18] Spearim, P. 'Garruu (Uncle) Paul Spearim (Winanga-li Gii) of the Gamilaraay Nation Tells the Story of Guwaymadhan, The Dark Blood Group', *The National Indigenous Times*, vol. 13, no. 344, p. 18, 2014.

[19] Testart, A. 'Some puzzling dualistic classifications in New South Wales', *Bijdragen Tot de Taal-, Land- En Volkenkunde*, vol. 136, no. 1, pp. 64–89, 1980.

[20] Tindale, N.B. et al. *Aboriginal Tribes of Australia: their terrain, environmental controls, distribution, limits and proper names.* Australian National University Press, Canberra, 1974.

[21] Wafer, J.W. et al. *A Handbook of Aboriginal languages of New South Wales and the Australian Capital Territory*. Muurrbay Aboriginal Language and Culture Cooperative, Nambucca Heads, 2008.

[22] Walker, K. *The Dawn Is at Hand*. Jacaranda Press, Brisbane, 1966.

[23] Waters, J. 'Experimenting with an Indigenous Model of Time', 2021. Accessed: 2022-10-10, <baayangali.com/blog/experimenting-with-an-indigenous-model-of-time>.

[24] Waters, J. 'Fractal Science and Indigenous Governance', 2021. Accessed: 2022-10-26, <baayangali.com/blog/fractal-science-and-indigenous-governance>.

[25] Whittaker, A. 'Not My Problem: on the Colonial Fantasy', *Sydney Review of Books*, 2019. Accessed: 2023-05-02.